Holt

ELEMENTARY SCIENCE

Joseph Abruscato
Joan Wade Fossaceca
Jack Hassard
Donald Peck

HOLT, RINEHART AND WINSTON, PUBLISHERS
New York • Toronto • London • Sydney

THE AUTHORS

Joseph Abruscato
Associate Dean
College of Education and Social Services
University of Vermont
Burlington, Vermont

Joan Wade Fossaceca
Teacher
Pointview Elementary School
Westerville City Schools
Westerville, Ohio

Jack Hassard
Professor
College of Education
Georgia State University
Atlanta, Georgia

Donald Peck
Supervisor of Science
Woodbridge Township School District
Woodbridge, New Jersey

Editorial Development William N. Moore, Lois Eskin, Roger R. Rogalin, Stephanie S. Blank
Editorial Processing Margaret M. Byrne, Regina Chilcoat, Shelley L. Feiler
Art and Production Frank P. Lamacchia, Vivian Fenster, Fred C. Pusterla, Robin Swenson, Russell Dian, Paula Darmofal, Theresa A. Harmsen, Susan Gombocz, Fran Conte, Dorina M. Virdo
Product Managers Ronald E. Suchodolski, John W. M. Cooke
Advisory Board John Boynton, Max Callahan, Glenn Hartman, Norman Hughes, Albert LeFevre, Douglas Nash, Jon Permar, Dennis Spurgeon
Consultant John Matejowsky
Researchers Eileen Kelly, Pamela Floch, Gerard LaVan

ART CREDITS

Robert Alcorn, pages 144, 145, 149, 153, 160
Howard Berelson, pages 96, 103, 134
Lisa Bonforte, pages 115, 124, 127, 128, 133
Marie De John, pages 6, 14, 76, 106
Bert Dodson, pages 16, 49, 58, 66,
Heidi Palmer, pages 10, 33, 35, 70, 79, 84-85, 107
Jerry Zimmerman, pages 9, 167

Photo acknowledgments appear on page 168.
Cover photograph by Steve Lissau.

Printed in the United States of America
ISBN 0-03-040481-9
01234 071 9876543

THE CONSULTANTS

Content Consultants

Edward E. C. Clebsch, Ph.D.
Professor of Botany
University of Tennessee
Knoxville, Tennessee

Jerry Faughn, Ph.D.
Professor of Physics
Eastern Kentucky University
Richmond, Kentucky

Ellen M. Herron, Ph.D.
Assistant Director
Lamont-Doherty Geological Observatory
Palisades, New York

Margaret A. LeMone, Ph.D.
Scientist
National Center for Atmospheric Research
Boulder, Colorado

W. T. Lippincott, Ph.D.
Professor of Chemistry
University of Arizona
Tucson, Arizona

Gary Peterson, D.A.
Assoc. Professor of Biology
South Dakota State University
Brookings, South Dakota

Arne E. Slettebak
Professor of Astronomy
Ohio State University
Columbus, Ohio

Gordon Taylor
Principal
Estey School
Saskatoon, Saskatchewan

Teacher Consultants

Peggy Ann Archacki
Assistant Supervisor of Science
Cleveland Public Schools
Cleveland, Ohio

Thomas L. Beck
Teacher
Evening Street Elementary School
Worthington, Ohio

Lynn T. Cluff
Teacher
Central School
South Burlington, Vermont

Linda Coffey, Ph.D.
Director of Early Childhood/Ed.
Broward County Public Schools
Broward County, Florida

Marcia Lambek
Teacher
School One
Scotch Plains, New Jersey

Carole Rutland
Teacher
Muscogee County Public Schools
Columbus, Georgia

Judy Woodward
Science Advisor
Delevan Science Center
Los Angeles City Schools
Los Angeles, California

Rina Zucker
Teacher
Ashford Elementary School
Houston, Texas

Reading Consultants

Paul Greenfield
Associate Professor
English and Humanities
Dutchess Community College
Poughkeepsie, New York

Judith Linscott Martin
Reading Specialist
Montgomery County Public Schools
Montgomery County, Maryland

Evelyn Mason
Elementary Language Arts Supervisor
Indianapolis Public Schools
Indianapolis, Indiana

PILOT SCHOOLS

We gratefully acknowledge the help of the teachers and students who field-tested portions of the Holt Elementary Science program in the spring of 1977. Their comments and criticisms were used to improve the program. The field teachers were:

Margaret Rodriquez
Encinal School
Morgan Hill, CA

Viola Sando
Murphy School
Stamford, CT

Lenore Ambrose
Palmetto El Sch
Miami, FL

Judy Kaplan
Lake Stevens El Sch
OpaLock, FL

Beth Williams
Laura Childs El Sch
Bloomington, IN

David Allen
Howard C. Reiche Sch
Portland, ME

Stephanie Barnhart
Immac Heart of Mary Sch
Towson, MD

John Cooney
Mill Swan El Sch
Worcester, MA

Eileen Martin
Stearns School
Pittsfield, MA

Phil Maines
Fountain El Sch
Grand Rapids, MI

Olivia McKinney
Vernor El Sch
Detroit, MI

Ellen Stob
Hillcrest El Sch
Grand Rapids, MI

Dennis Davis
Portland El Sch
Richfield, MN

Gary Hawkins
Cambridge El Sch
Cambridge, MN

M. Foster
Green Trails Sch
Chesterfield, MO

A. J. Sullivan
Ferguson Florissant Sch Dist
Ferguson, MO

Rick Ashworth
Choteau School
Choteau, MT

Nancy Ritter
Fort Benton Sch
Ft. Benton, MT

J. Ely
Burgess School
Berlin, NH

Edward Douglas
Sumner El Sch
Camden, NJ

Nancy Hearst
Pennypacher El Sch
Willingboro, NJ

Lee Ferrera
St. Jerome Sch
Brooklyn, NY

Robert Kase
Public School 178
Jamaica, NY

Lois Parker
Bloomfield El Sch
Holcomb, NY

Lorraine Sharp
95th Street School
Niagara Falls, NY

Paul Snyder
93rd Street School
Niagara Falls, NY

Helen Suchy
Theo Roosevelt El Sch
Binghamton, NY

Edward Wianecki
School 43
Buffalo, NY

Patricia Brazas
Peck El Sch
Greensboro, NC

Ann Schwabeland
Irving Park El Sch
Greensboro, NC

Bernice Perry
Brooks El Sch
Raleigh, NC

John Foley
Lewis and Clark Sch
Fargo, ND

Manna Hay
Arlington Sch
Toledo, OH

Charles Knepshield
Taft El Sch
Middletown, OH

Dawn Rowe
Brush El Sch
Grafton, OH

Ross Neidich
West Branch El Sch
Bradford, PA

Frank Rice, Jr.
Westview El Sch
Spartanburg, SC

Wilma Todd
Cooper El Sch
Garland, TX

James Goodwin
East Salem Sch
Salem, VA

Dean Rickabaugh
Highland Park Sch
Roanoke, VA

Patricia Kinner
St. Aemilian's Sch
Milwaukee, WI

Kathryn Lee
Falk El Sch
Madison, WI

Many thanks also to the principals, supervisors, and science coordinators who assisted in the arrangements for the field test.

Contents

UNIT 3

How's The Weather? 54

UNIT 4

Oceans And Beaches 82

UNIT 5

All Kinds Of Animals 110

UNIT 6

Plants 138

We hear many sounds every day.
We see lights all around us.
What sounds and lights are in these pictures?

UNIT

1 LOOK AND LISTEN

1 WHAT IS SOUND?

A **sound** is made when something moves back and forth very fast.

These strings are moving fast.

How are these children making sounds?

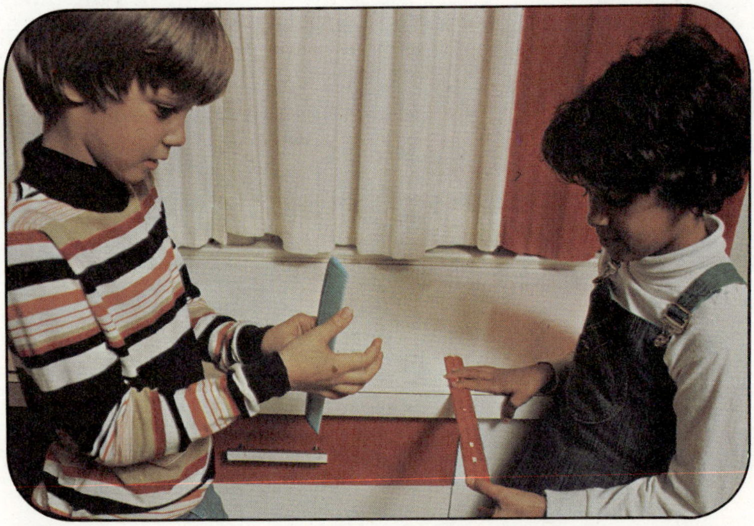

This girl tapped a glass filled with water.

The glass, the water, and the air inside the glass moved fast.

This made a sound.

Tap on a desk.

Do you see anything move back and forth fast?

Sometimes things move back and forth so fast that you do not see them move.

Many sounds are made when air moves back and forth fast.

Blowing into a bottle makes the air inside move fast.

This makes a sound.

How are these children making sounds?

ACTIVITY

1. Stretch a rubber band around a ruler.
2. Slide a pencil under the rubber band.
3. Pluck the rubber band.
4. What do you see?
   What do you hear?
5. How many different sounds can you make?

2 SOUNDS WE HEAR

Sounds are all around us.
There are many kinds of sounds.
Look at the picture.
What sounds are being made?

Some sounds are **high.**

A whistle is a high sound.

What high sounds do you know?

Some sounds are **low.**

A cow's moo is a low sound.

What low sounds do you know?

Some sounds are **loud.**

A fire truck's horn makes a loud sound.

What loud sounds do you know?

Some sounds are **soft.**

A whisper is a soft sound.

What soft sounds do you know?

Some sounds are too loud for us.
Very loud sounds can hurt out ears.
What loud sounds hurt your ears?

3 SOUND MAKERS

Many things make sounds.
People make sounds.
Machines make sounds.
Things in **nature** make sounds.

Some sounds make music.
What things are these children using to make music?

Animals make many sounds.

The meow of a kitten is soft.

The roar of a lion is loud.

What other sounds do animals make?

ACTIVITY

You are a sound maker too.

1. Put your hand on your throat.

2. Talk to a friend.

3. What do you feel?

4. How many different sounds can you make?

4 LIGHT

The sun gives us **light** in the day.
The moon and stars give us light at night.
We cannot see without light.

We get most of our light from the sun.
But other things give us light too.
What things give us light?

Light moves in a straight line.

Light bounces off things.

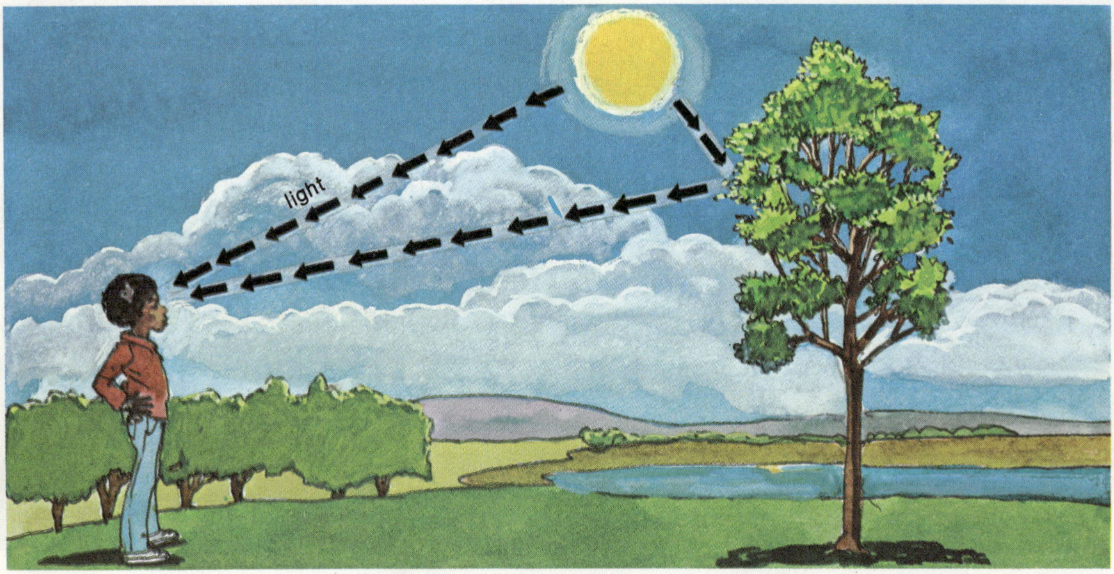

When light goes into our eyes, we see.
The girl sees the sun when its light goes into her eyes.
The girl sees the tree when light from the sun bounces off the tree into her eyes.

ACTIVITY

1. Roll a piece of paper into a tube.
2. Look at a friend through the tube.
3. Bend the paper and look again.
4. Can you see your friend?

5 LETTING LIGHT IN

The boy is looking out the window.
He can see through the glass.
Light passes through the glass.

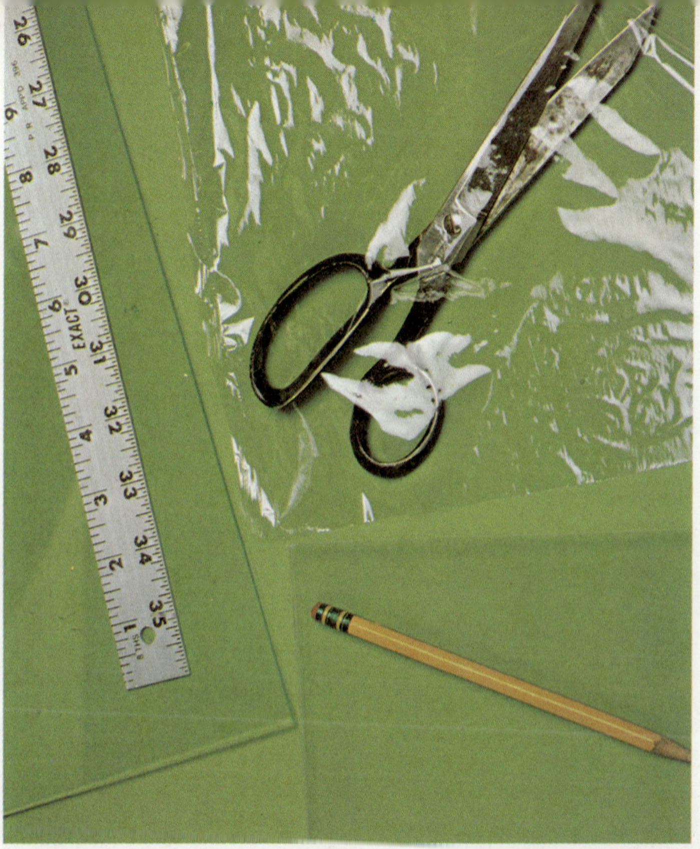

All light passes
through these things.

We can see through
these things.

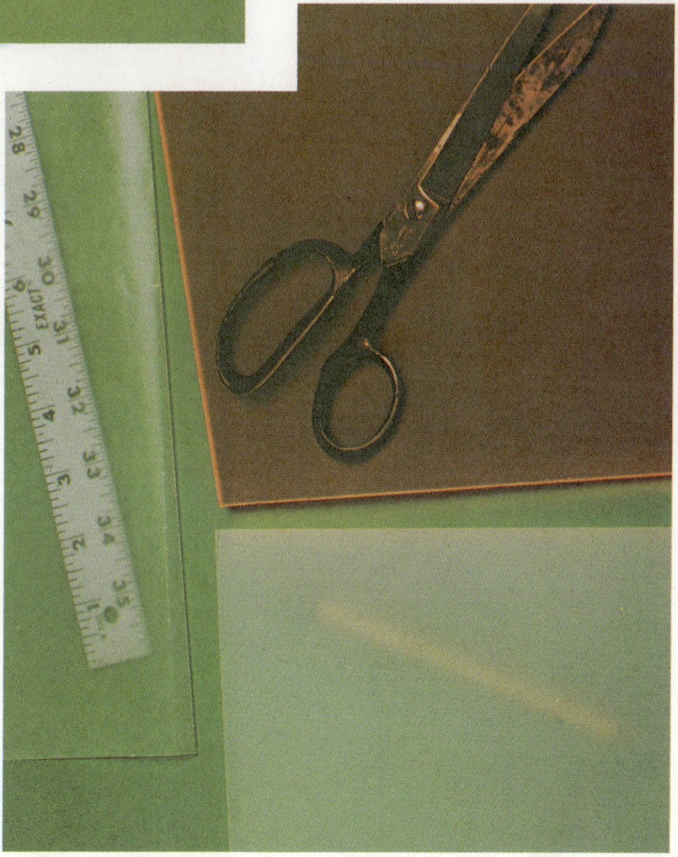

Some light passes
through these things.

We cannot see clearly
through these things.

No light passes through these things.
We cannot see through these things.

What things in this room let all light through?
Let some light through?
Let no light through?

ACTIVITY

1. Hold some things up to a light.
2. How well can you see a pencil through these things?
3. How much light passes through?
4. Mark your answers on a chart.

6 SHADOWS

The sun is shining in the sky.
There are **shadows** all around.
What things have shadows?

A shadow is made when light cannot pass through something.
The boy is making a shadow on the wall.
What is he using to make the shadow?

The girl and the tree both have shadows.
Where is the boy's shadow?

Shadows point away from the light.

Where is the light?
How do you know?

Where is the sun?
How do you know?

The size of shadows can change.

This shadow is short.

This shadow is long.

What makes shadows change size?

Look at each picture.
Is the sun high or low in the sky?
Can you tell where the sun is by
looking at the shadows?

MAIN IDEAS

- A sound is made when something moves back and forth very fast.
- Sounds can be high or low, soft or loud.
- We need light to see.
- Light bounces off many things.
- Some things give off their own light.
- Light travels in a straight line.
- More light passes through some things than others.
- A shadow is made when something does not let light pass through it.

QUESTIONS

1. How is sound made?
2. Name a loud sound, soft sound, high sound, and low sound.
3. Name four things that give us light.
4. How does light travel?
5. Name two things that all light passes through.
6. Name two things that some light passes through.
7. Name two things that no light passes through.
8. How is a shadow made?

SOMETHING EXTRA

This instrument is called a steel drum. Each mark on the drum is a different note.

When the man hits the drum, the top of the drum moves back and forth very fast.

It makes a beautiful sound.

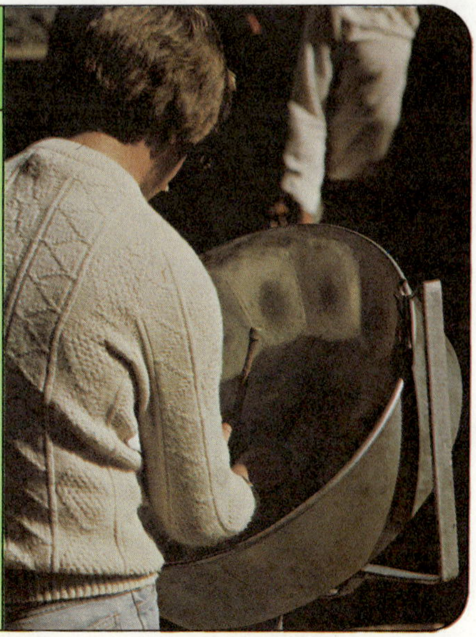

We move things every day.
We try to make some things go fast.
We try to make other things go slow.
What is moving here?

UNIT

2 MOTION

1 WHAT IS A FORCE?

We can **push** things.
We can **pull** things.

This is a push.

This is a pull.

A **force** is a push or a pull.

What kind of force is this?

What kind of force is this?

A **strong** force is a big push or pull.
This big engine is pulling the train.
It pulls with a strong force.

A **weak** force is a small push or pull.
This small engine is pulling the train.
It pulls with a weak force.

ACTIVITY

Do		Is It A Push?	Is It A Pull?
1. Move your chair under your desk.			
2. Open a door.			
3. Close a door.			
4. Open a drawer.			
5. Open a book.			
6. Erase the chalkboard.			

1. Do each thing.
2. Is it a push or a pull?
3. Write your answers on a chart.

2 MOVING THINGS

Things cannot move by themselves.
Forces make things move.

A push is making
this ball move.

A pull is making
this toy move.

What forces are making these things move?

A strong force is needed to move **heavy** things.

A weak force is needed to move **light** things.

Which wagon needs more force to be moved?

How do you know?

Forces are needed to move things up or down.

More force is needed to move something up a hill than down a hill.

3 CHANGING THINGS

Forces can change the way things look.
Forces can change the **size** of things.
Forces can change the **shape** of things.

Forces can change the **direction** things move.

The girl changed the direction of the ball.

What force did she use?

What force does the truck use to change its direction?

A force can change the **speed** of moving things.

A force can make things move faster. The more force the girl uses, the faster she will go.

A force can make things go slower. The more force the boy uses, the slower the wagon will go.

ACTIVITY

1. Make a clay ball flat.
2. Make this clay long.

3. Make a clay ball move fast.
4. Make a clay ball move slowly.

5. What forces did you use to change the clay?

4 SLOWING THINGS DOWN

This boy cannot move fast.

The wheels of his bicycle are **rubbing** against the grass.

This slows the bicycle down.

When something moves, it rubs against other things.

This rubbing slows moving things down.

The truck is moving.
What will make it go slower?

Everything that moves rubs against air.

Air can slow things.

Moving on something rough slows things.
Which child moves slower?
Which child needs more force to move?
Why?

5 MOVING SLOWLY

The girl is riding her scooter.
She sees a car.
She wants to stop.
She drags her foot on the ground.
How does this help her stop?

Sometimes we want things to move slower.

Sand and stones are put on roads to make them rough.

This helps cars go slower.

Some people put snow tires on their cars.

How does this help the car move slower?

Many shoes are made to help keep you from slipping.

They help you slow down.

Look at the bottoms of these shoes.

How would they keep you from slipping?

6 MAKING THINGS MOVE EASILY

Sometimes we want things to move faster and easier.

It is faster and easier to move on smooth things.

What is smooth here?

Wheels make it easier to move things.

Many things have wheels.
How many wheels does each thing have?

The girl is putting **oil** on her toy.
The oil makes the toy's parts move smoothly against each other.
This makes the toy move easily.

ACTIVITY

1. Rub two fingers together.
2. Put some oil on your fingers.
3. Rub your fingers together again.
4. What did the oil do?

MAIN IDEAS

- A force is a push or a pull.
- A force is needed to move things.
- A force can change the size and shape of things.
- A force can change the speed or direction of things.
- Moving over rough places slows things down.
- Wheels and oil help make things move more easily.

QUESTIONS

1. What is a force?
2. What makes things move?
3. What can forces do?
4. Is it easier to move a heavy object or a light object?
5. What is slowing the truck in picture 3?
6. How can people move safely on icy roads?
7. Name two things that make moving easier.

SOMETHING EXTRA

This lady is putting oil into her car.

A car engine has many moving parts.

The parts rub against each other as they move.

The oil makes the parts of the engine move more easily.

This keeps the parts from wearing out.

Weather is all around us.
Weather helps us plan our days.
What kind of weather do you see in these pictures?

UNIT

3 HOW'S THE WEATHER?

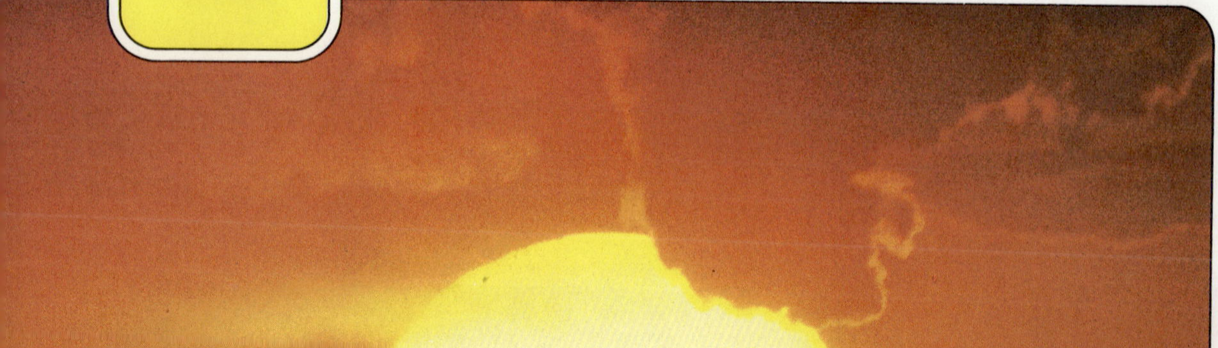

1 HOT AND COLD

Weather can change from day to day.

Weather can change from place to place.

Some places are hot.

It is hot here.

How do you know?

Some places are cold.

It is cold here.

How do you know?

The sun **heats** the earth.

Temperature tells you how hot or cold something is.

We measure temperature by using a thermometer.

What are the children doing?

Where is it hotter in this picture?

The line inside the thermometer gets longer as the temperature gets hotter.

Look at this thermometer.

Is the temperature hot or cold?

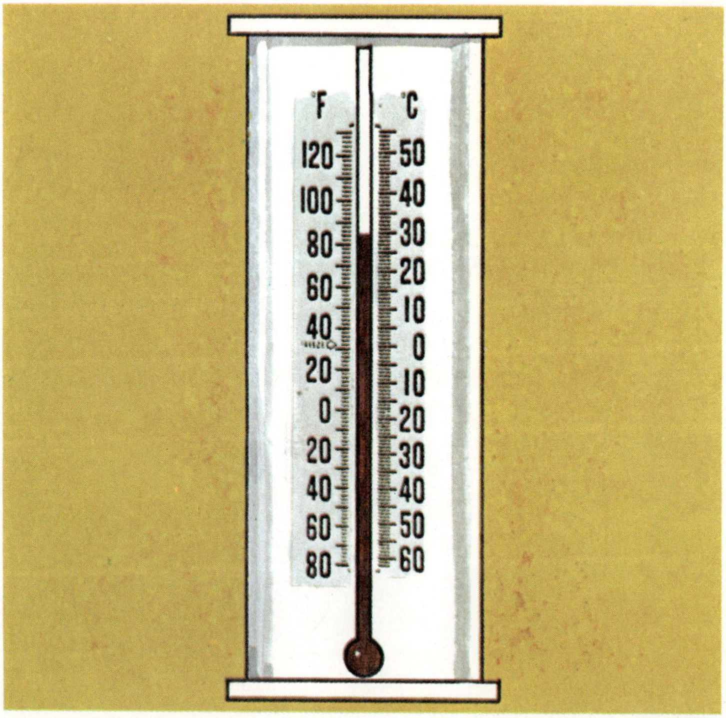

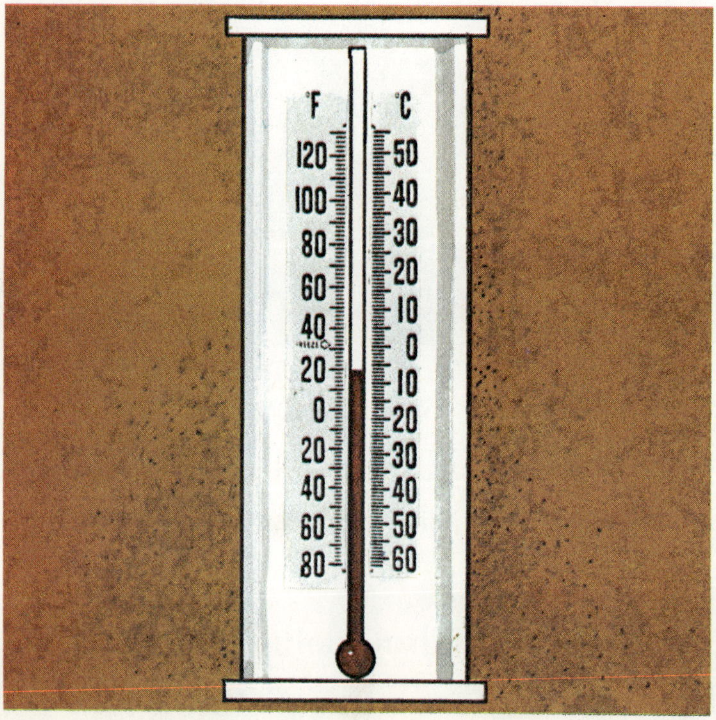

The line inside the thermometer gets shorter as the temperature gets colder.

Look at this thermometer.

Is the temperature hot or cold?

ACTIVITY

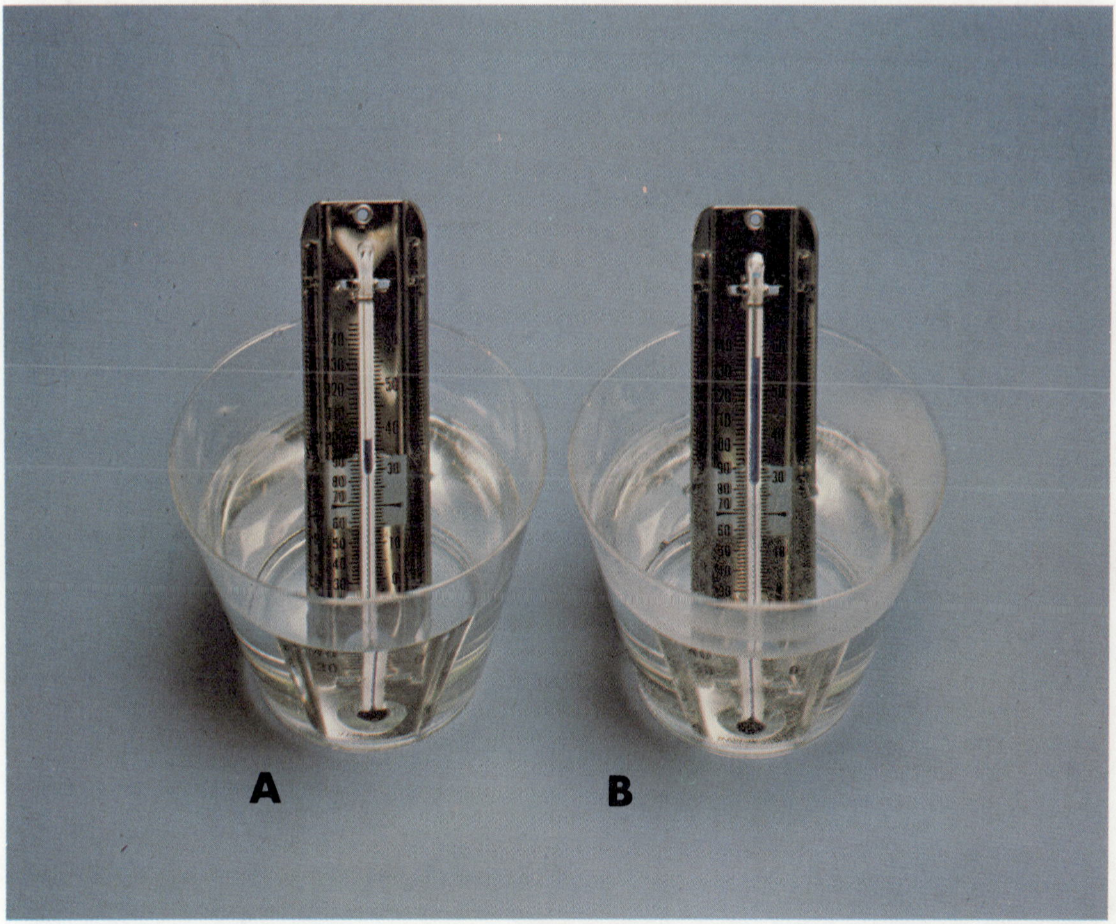

1. Place a thermometer into warm water.
2. Place a thermometer into cold water.
3. Watch the lines inside the thermometers.
4. In which direction does the line move in each thermometer?

2 LOOKING AT THE WEATHER

The sun, the air, and water help make the weather.

There are many kinds of weather.

What are these children doing?

The sun is shining.
It is a sunny day.

Moving air is called **wind.**

The wind can move things.

The children are finding out which direction the wind is coming from.

How can they do this?

The sky is covered with **clouds.**

The sun is behind the clouds.

The sun cannot be seen.

It is a cloudy day.

Have you ever walked through a cloud?

A low cloud on or near the ground is called **fog.**

What kind of day is it in this picture?

Rain is falling on the ground.

What does it look like outside on a rainy day?

It is cold outside.

Snow is on the ground.

Snow falls in pretty shapes called **snowflakes.**

3 WHERE DOES WATER GO?

Joey sailed his boat in a puddle.

When he came back the next day the puddle was gone.

Where did the water go?

The water in the puddle dried up.

It went into the air.

What made the water go into the air?

Heat helps make water go into the air.

Where did heat come from to dry up the ground?

The wind helps make water go into the air too.

What is helping to dry these clothes?

Water that is in the air is called **water vapor.**

What is making the water in the pot change into water vapor?

Water goes into the air in many places. Can you name some of these places?

ACTIVITY

1. Fill some jars with water.
2. Put them in different places in the room.
3. What will happen to the water in the jars?

4 WHERE DOES WATER COME FROM?

We cannot see water vapor but it is always in the air.

Water vapor can change back into water when the air cools.

The air is very cold high above the earth.

In cold air the water vapor changes into drops of water.

The drops of water and dust in the air make clouds.

When the drops of water get too heavy to float in the air, they begin to fall.

Then we have rain.

Heat from the sun changes water into water vapor.

The water vapor goes into the air.

Cold air changes water vapor back to water.

The drops of water mix with dust to make clouds.

What do you think will happen next?

ACTIVITY

We can make rain in the classroom.

1. Look at this picture.
2. What is making the water change into water vapor?
3. What is making the water vapor turn into water?
4. When will the drops on the bottom of the pan begin to fall?

5 CLOUDS

There are many different kinds of clouds.

These clouds cover the sky.

They are dark because they are holding a lot of water.

They are rain clouds.

These clouds look like cotton balls.

They seem to be always moving and changing shape.

The sun peeks through these clouds.

How are these clouds different from rain clouds?

You can see these clouds on a clear sunny day.

They are the highest clouds in the sky.

They are so thin you can almost see through them.

They look like feathers in the sky.

ACTIVITY

1. Make a cloud chart.
2. Use cotton to show the different kinds of clouds.
3. Draw a picture under the clouds to show the weather each kind of cloud brings.

6 WEATHER CHANGES

Weather changes from season to season.

In some places the weather changes from very hot in the summer to very cold in the winter.

In other places the weather changes from very wet in the summer to very dry in the winter.

How does the weather change where you live?

Plants change as the weather and seasons change.

The leaves of some plants change color in the fall.

Some plants lose their leaves in the winter.

Many flowers grow in the spring.

Animals behave differently as the weather and seasons change.

Some animals sleep through the winter.

They wake up in the spring.

In the winter some animals move to places where the weather is warm.

Many animals have babies in the spring.

In the winter daytime is shorter than nighttime.

In places where it is cold, people wear heavy clothes to keep themselves warm.

In the summer daytime is longer than nighttime.

People wear light clothes to keep themselves cool in the summer.

MAIN IDEAS

- Weather can change from day to day and from place to place.
- A thermometer measures temperature.
- There are many kinds of weather.
- Heat and wind change water into water vapor.
- Cool air temperatures change water vapor into water.
- There are many kinds of clouds.
- The weather and seasons cause many changes.

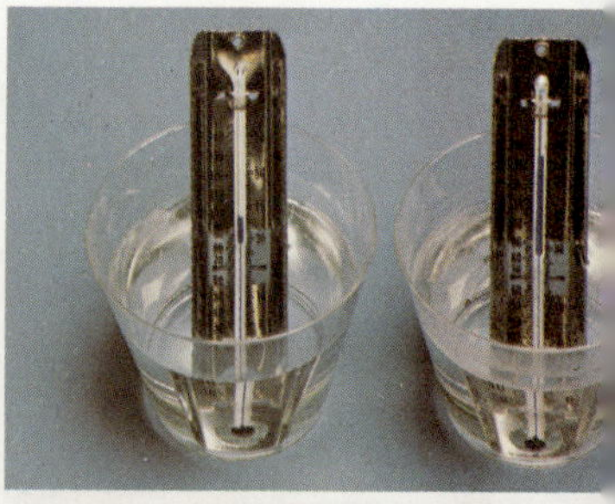

QUESTIONS

1. Do all places have the same weather?
2. How does a thermometer measure temperature?
3. What is fog?
4. What is water vapor?
5. What makes water change into water vapor?
6. What is a cloud made of?
7. What happens to people, plants, and animals when the weather and seasons change?

SOMETHING EXTRA

These people study the weather. They know many things about clouds, wind, and rain.

They use maps, pictures, and weather instruments to help them with their work.

They try to tell us what the weather will be.

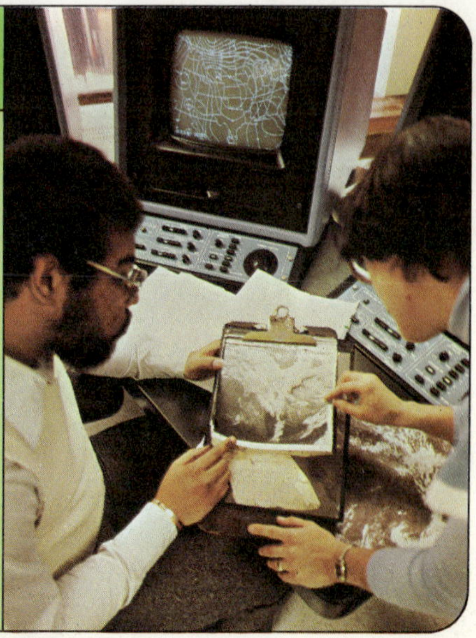

Many people enjoy the oceans and beaches. Many plants and animals live in the oceans and on the beaches.

How are these people, plants, and animals using the oceans and beaches?

UNIT

4 OCEANS AND BEACHES

1 OUR OCEANS AND BEACHES

This is a map of the **earth.**
There is more water than land on earth.

Some of the water is in **ponds, lakes,** and **rivers.**
Most of the water is in the **oceans.**

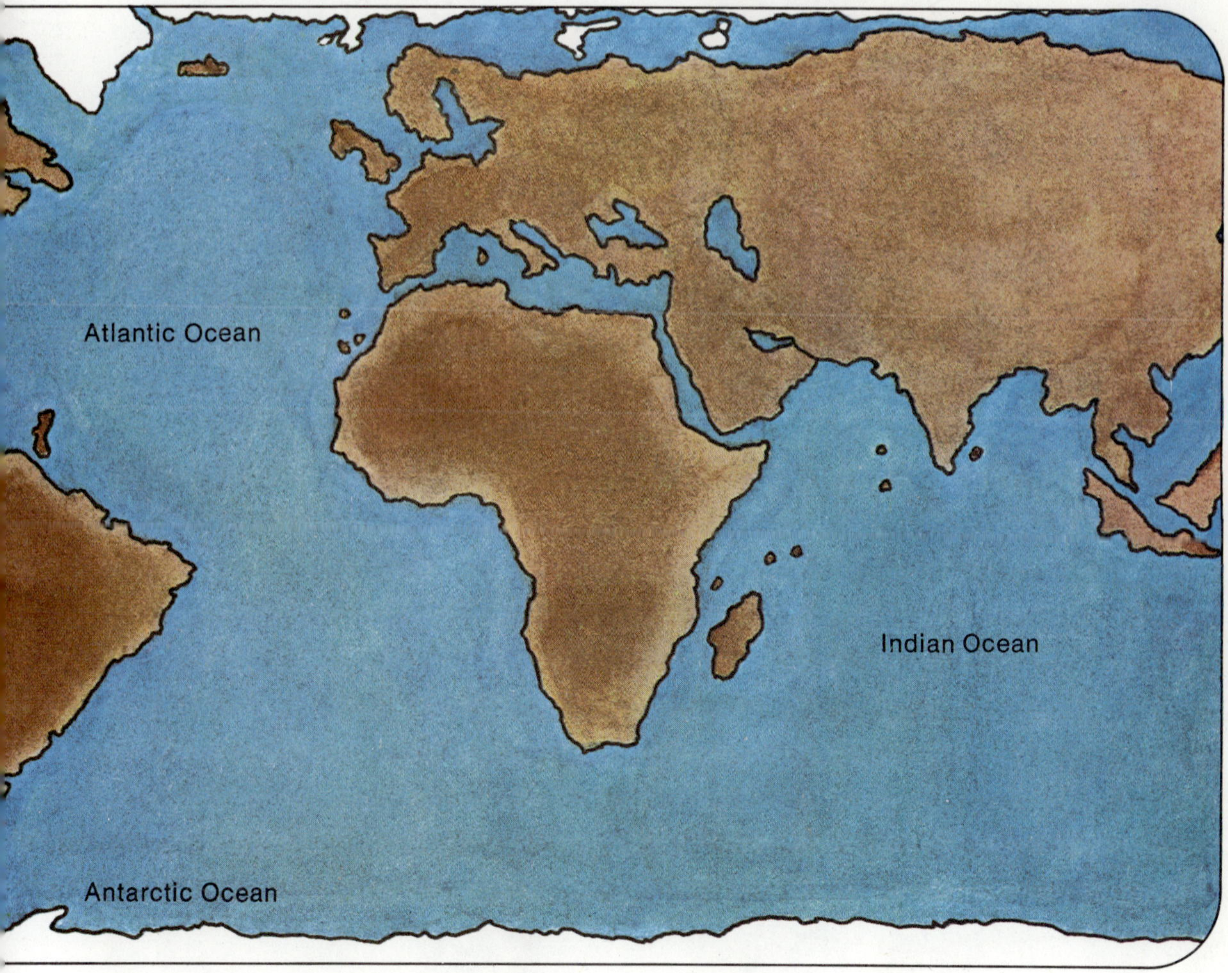

The earth has five oceans.
Oceans are **shallow,** or not very deep,
in some places.
They are **deep** in other places.
Ocean water can be warm or cold.

Beaches can be many different shapes and sizes.

There are many kinds of beaches.

There are **rocky** beaches.

There are **sandy** beaches.

There are beaches made out of crushed **shells**.

ACTIVITY

1. Make a model of a beach.
2. Use sand or pebbles, water, and a tray.

2 HOW ARE BEACHES MADE?

Beaches are found near water.

There are beaches near some lakes.

There are beaches near some rivers.

Is there a beach near where you live?

Water is rubbing sand against these rocks.

When this happens, the rocks will slowly break into sand.

The sand at this beach was made a long time ago from very large rocks.

What helped change the rocks into sand?

The wind blows sand against the rocks.

The wind and sand **scrape** the rocks.

This changes the rocks into sand.

Wind and water work together to help make beaches.

ACTIVITY

1. Rub two rocks together over a paper towel.
2. What falls on the towel?

3 OCEAN WATER

Water in the ocean is different from water we drink.

Ocean water is different from water in most ponds, lakes, and rivers.

Ocean water is **salty.**

The water in the ocean is always moving.

Waves are made when the wind moves the water.

At certain times each day, the ocean water moves away from the beach.

When this happens, it is called **low tide.**

At other times each day, the ocean water moves closer to the beach.

When this happens, it is called **high tide.**

The tides, waves, and wind move sand.

This slowly changes the way beaches look.

A **storm** has quickly changed the way this beach looks.

What did the storm do?

ACTIVITY

1. Make a beach with sand or pebbles, water, and a tray.
2. Tilt the pan to show high tide.

3. Tilt the pan to show low tide.
4. How do high tides and low tides change your beach?

4 OCEAN PLANTS

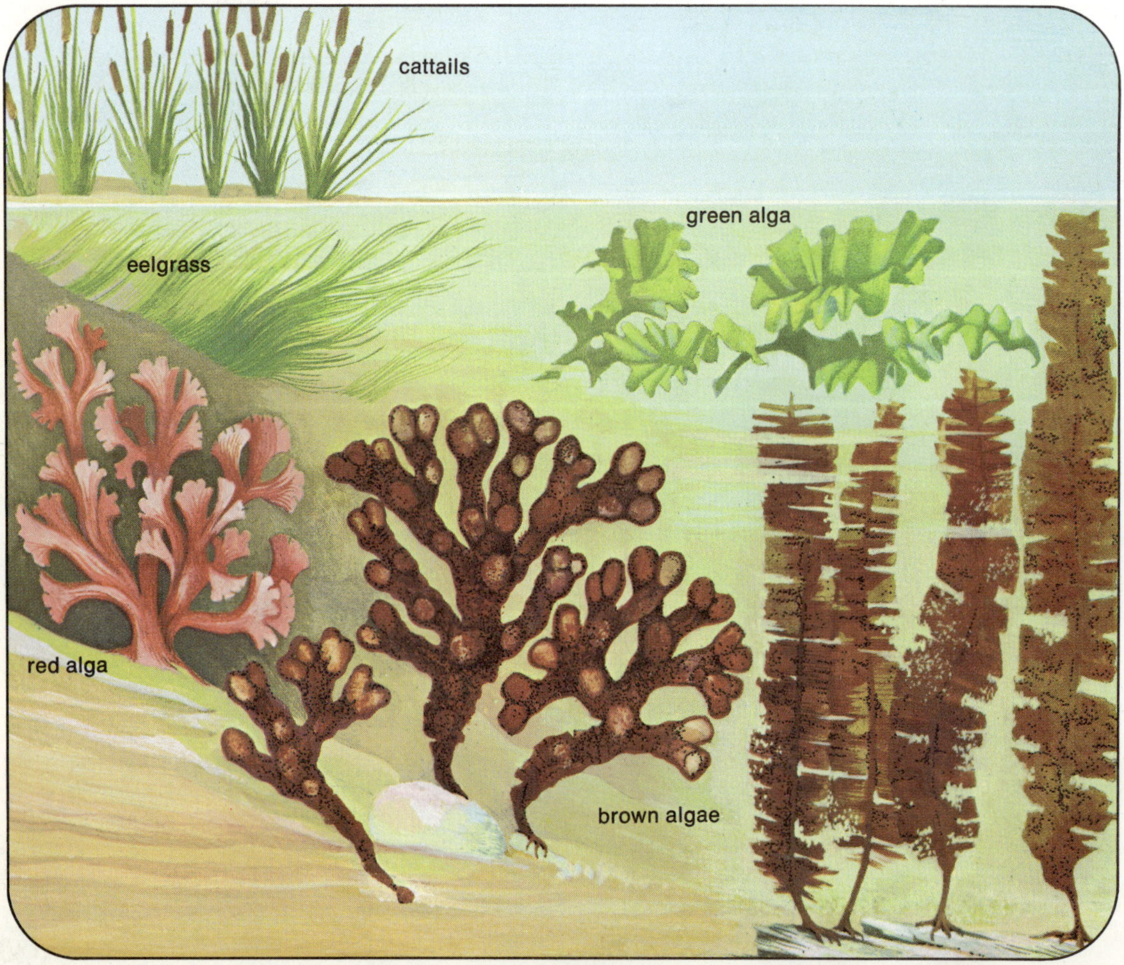

Different kinds of plants can be found near beaches and in ocean water.

Some of these plants grow on the ocean bottom.

Some cling to rocks.

Some float on water.

Plants need sunlight to live.

They can only grow where they can get sunlight.

Can plants grow in very deep water?

Many of the plants in the ocean are very small.

They are so small you cannot see them without a **microscope.**

A microscope makes things look bigger.

This is what these small plants look like through a microscope.

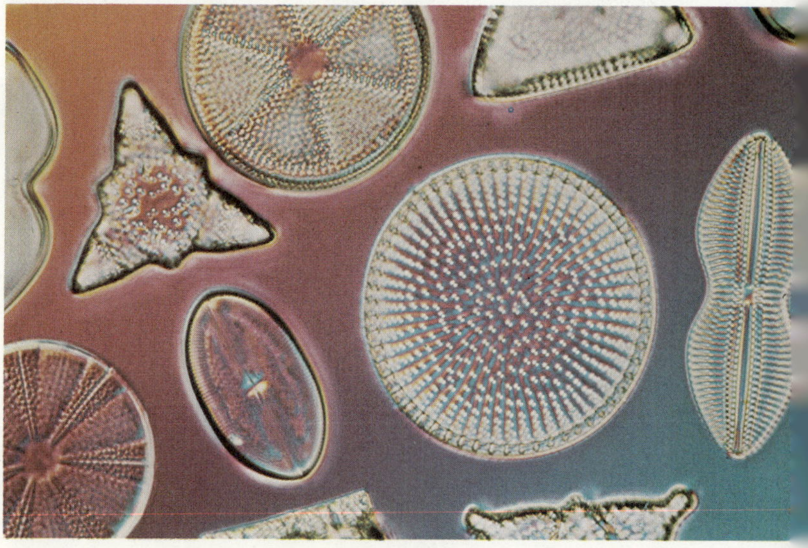

These small plants float near the top of the water.

Many animals in the ocean use these plants for food.

5 OCEAN ANIMALS

There are many kinds of animals living in or near the ocean.

Some of the animals **swim** or **float** in the water.

Some **crawl** on the bottom.

Others live on the beach near the water.

The blue whale is the biggest animal on earth.

It lives in the ocean.

Some of the smallest animals live in the ocean too.

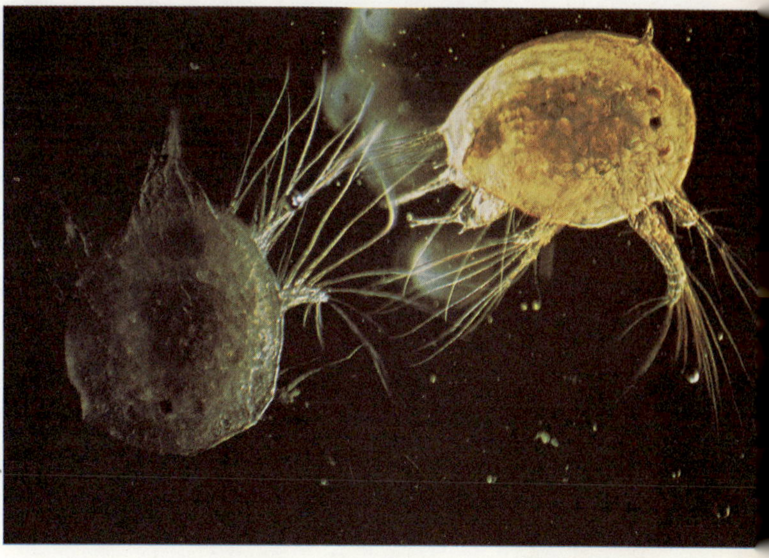

This is a **sponge.**

It is an animal.

It gets its food from the water that passes through its holes.

This is **coral.**

Coral is a sea animal.

It does not move from place to place.

It gets its food from the water.

Animals make their homes in many different places in the ocean.

What animals do you see in the picture?

Do they all live in the same part of the ocean?

6 THE OCEAN AND LIVING THINGS

All plants and animals need food to stay alive.

Green plants use sunlight to make food.

Many animals use green plants for food.

These fish eat green plants.

This big fish eats the small fish for food. What would happen to the big fish if all the small fish were gone?

People eat fish for food.

Animals need plants and other animals for food.
This is called a **food chain.**

People use the oceans for many things.

They use it for food, for travel, and for fun.

People can **misuse** the oceans too.

Sometimes people do things to the ocean that kill many plants and animals.

What does this do to the food chain?

ACTIVITY

1. Which picture shows the beginning of a food chain?
2. Which picture shows the end?
3. Put these pictures in order.

MAIN IDEAS

- Most water on earth is found in the oceans.
- Beaches are found near water.
- Wind and water help make and change beaches.
- There are many living things in and near the ocean.
- Animals and plants of the ocean need each other for food.
- People use the oceans in many ways.

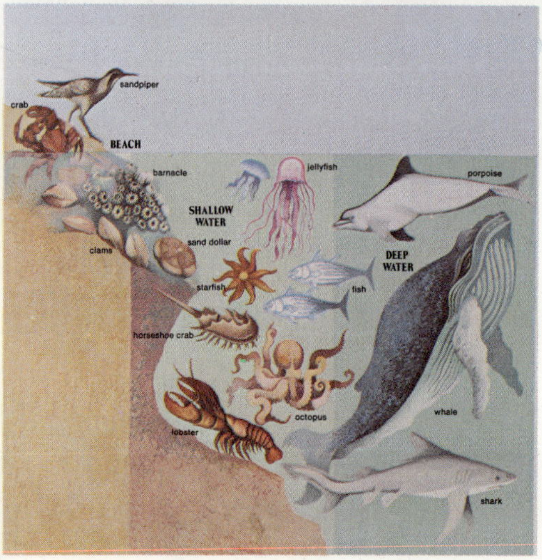

QUESTIONS

1. Where is most of the water on earth?
2. Do all beaches look the same?
3. What makes beaches?
4. What makes waves?
5. Name two places where ocean plants live.
6. How do ocean animals move?
7. What is a food chain?
8. How do people use the oceans?

SOMETHING EXTRA

People are finding new ways to use the oceans.

Some people are starting ocean farms.

They are growing plants and raising fish, clams, and oysters.

Someday much of our food may come from the farms of the ocean.

Many different animals live around us.
What animals do you see?
What are they doing?
How do you know that they are alive?

UNIT

5 ALL KINDS OF ANIMALS

1 MAMMALS

Mammals are animals that have **hair** or **fur** on their bodies.

Mammals are **warm-blooded.**

Their body temperature always stays the same.

These animals are mammals.

How do you know?

Mammals are born alive.

The babies drink milk from their mother.

Mammals breathe in air through their **lungs.**

These animals are mammals. How do you know?

Mammals live
in many different places.

They eat many kinds
of foods.

Sheep live on the
grassland.

They eat only plants.

Lions live in
the **jungle.**

They eat only
meat.

Bears live in the **forest.**

They eat both plants
and animals.

ACTIVITY

1. snake

2. bear

3. frog

4. robin

5. deer

6. alligator

7. cat

8. dog

9. goldfish

10. lizard

11. squirrel

12. monkey

13. turtle

14. clownfish

15. owl

1. Which of these animals are mammals?

2. How do you know?

2 BIRDS

Birds are animals that have **wings** and **feathers.**

Birds can be many different colors.

What colors are these birds?

Feathers help keep birds warm and dry.

Birds use their wings and feathers to **fly.**

Birds use their **beaks** to get food.

Some birds have long beaks.

Some birds have short beaks.

Birds use their feet to do many things.

A duck uses its feet to swim.

This bird uses its feet to catch food.

What other things do birds do with their feet?

Some birds make their **nests** in trees.

The mother bird lays her **eggs** in the nest.

She sits on them to keep them warm.

Baby birds will **hatch** from the eggs.

Birds are warm-blooded animals.
They breathe in air through their lungs.
Birds take care of their babies.
How are birds different from mammals?

ACTIVITY

1. Look at a feather of a bird.
2. Feel it. How does it feel?
3. Drop some water on the feather.
4. Does the water go through the feather?
5. How do feathers help a bird?

3 FISH

Fish are animals that live in water.

Some fish live in **fresh** water.

Some fish live in **salt** water.

A fish uses its **tail** and **fins** to swim.

Do you see the fins on these fish?

Fish are **cold-blooded** animals.
Their body temperature does not stay the same.
Their body temperature changes as the temperature around them changes.

All fish have **gills.**

A fish uses its gills to breathe.

The gills are in the slits on the side of a fish's head.

Fish are covered by hard **scales.** How do scales protect the fish?

Most fish are born from eggs.
Fish do not take care of their young.
How are fish different from birds and mammals?

ACTIVITY

1. Look at a fish.
2. Find and count its fins.

3. Find the gills.
4. Find the scales.
5. Make a picture of a fish.
6. Label the parts of the fish.

4 AMPHIBIANS

Amphibians are animals that live on land or in water.
These animals are amphibians.
Can you name them?

Amphibians lay eggs in water.

The babies hatch from the eggs.

They have gills like fish.

The babies breathe and grow in water.

Amphibians grow lungs and breathe air as they get older.

Then they are able to live on land.

Amphibians are cold-blooded animals.

Most of them have wet, smooth skin.

How are amphibians different from other animals?

A baby frog is called a **tadpole.**

Tadpoles live in water.

Do tadpoles look like frogs?

A tadpole begins to look more like a frog as it gets older.
How does a tadpole change?

5 REPTILES

These animals are **reptiles.**
They live on land.
Their skin is dry, rough, and covered with scales.
How are these animals alike?
How are they different from each other?

Reptiles are cold-blooded.

They lay eggs on land.

The babies hatch from the eggs.

Many reptiles swim but they cannot breathe underwater.

Some reptiles eat **insects** and other small animals.

These two reptiles look almost alike. Can you see how they are different?

Many reptiles are the same color as the things around them.

Can you find the reptile in the picture?

Some reptiles use their **tongues** to help them **smell.**

Some reptiles **shed** their skins as they get larger.

There is new skin under the old skin.

6 DINOSAURS

A very long time ago animals called **dinosaurs** lived on earth.
These animals were reptiles.
They do not live today.
We learn about dinosaurs by looking at their bones.

This dinosaur was very big.

It lived in the water.

It ate only plants.

This dinosaur had very **sharp teeth.**

It lived on land.

It ate other animals.

This dinosaur also lived on land.
It ate only plants.
How did this dinosaur **protect** itself?

Fill in the chart about dinosaurs.

	Where did it live?	What did it eat?	How did it protect itself?

MAIN IDEAS

- There are many different kinds of animals.
- Animals can be placed into different groups.
- Animals in each group are the same in some ways.
- Animals grow and take care of themselves in different ways.
- Some animals that lived long ago do not live today.

QUESTIONS

1. How does a baby mammal get food?
2. What is the difference between a warm-blooded and cold-blooded animal?
3. How are birds and fish different from one another?
4. How does a fish breathe?
5. How is a tadpole different from a frog?
6. How are all reptiles the same?
7. What animal lived a long time ago?

SOMETHING EXTRA

Some animals can do things people cannot do.

This lizard just lost its tail.

A new tail will grow in its place.

Some people are trying to find out how the lizard can do this.

Plants are all around us.
There are many kinds of plants.
Plants grow in many places.
Do you know what plants need to grow?

UNIT

6 PLANTS

1 WHAT DO PLANTS NEED?

These green plants are living.
They need certain things to live.
What green plants do you see in this picture?
What do you think they need to live?

Green plants need **water** to live.

Water helps plants grow and stay healthy.

Which plant looks like it doesn't have enough water?

How do you know?

Most green plants need **soil** to live.

Soil holds water for the plants to use.

Soil has special things in it to help the plant stay healthy.

Green plants need **air** to live.

Some green plants use air that is in water.

All green plants need **sunlight** to live.

The sunlight helps green plants make food.

ACTIVITY

1. Take four plants and do this.
2. What do you think will happen to each plant?
3. Watch them for a few days.
4. What happened to each plant?

2 ROOTS

Most green plants have **roots.**
Most roots grow in soil or water.
Do the roots of these plants look the same?
How are they different?

Some plants grow many roots that spread out.

The roots hold the plant in the ground.

Some plants grow one thick root that grows deep into the soil.

A plant uses its roots to get water.
The water goes through the roots into the plant.

We eat the roots of many plants.
Do you know the names of these foods?
They are all roots of plants.

ACTIVITY

1. Grow potato roots like this.
2. Add some red color to the water.

3. Wait a few days.
4. Cut the potato in half.
5. What do you see?

3 THE PLANT STEM

The **stem** is the part of a plant between the roots and the leaves.

There are many kinds of stems.

The stem of this plant is soft and green.

The stem of this plant is hard and made of **wood.**

The **branches** of a plant are part of the stem.

The stem holds up the plant.

The stem holds the leaves up to the sunlight.

The water and food are carried through the stem to the plant.

We eat the stems of many plants.

Do you know the names of these foods?

They are all the stems of plants.

ACTIVITY

1. Put some celery in colored water.
2. Wait a day.

3. Cut a piece of the celery.
4. Look at the piece closely. What do you see?

4 THE LEAF

A **leaf** grows on the stem or branch of a plant.

Leaves can be many different shapes and sizes.

Look at these leaves.

How are they the same?

How are they different?

A green plant makes food in its leaves.

The leaf has tiny holes that let **water vapor** and air pass through.

This helps the leaf get what it needs to make food.

The leaves need sunlight, air, and water to make the food for the plant.

A green plant is the only living thing that can make its own food.

We eat the leaves of many plants.

Do you know the names of these foods?

They are all the leaves of plants.

ACTIVITY

1. Cover the leaf of a plant with clear jelly.

2. Look at the leaf after a few days.

3. What is happening to the leaf? Why?

5 FLOWERS AND FRUITS

Many plants have **flowers.**

There are many kinds of flowers.

How are these flowers the same?

How are they different?

Many flowers have a nice smell.
Do you have a favorite flower?

Flowers make **seeds.**

Many plants grow from seeds.

This flower will make a seed.

The seed will grow into a plant.

Seeds need water, air, and warm weather to grow.

What plant will grow from these seeds?

Fruits are made from flowers.

Fruits hold the seeds of the plant.

Some fruits are juicy.

Some fruits have have many seeds inside.

Some fruits have only one seed inside.

All these foods are fruits.
They all have seeds.
Can you name the fruits?

Some flowers make fruits that are hard and dry.

One kind of dry fruit is called a **nut.**

Nuts have seeds inside.

This nut is good to eat.
It is a **peanut.**
It comes from a peanut plant.

An **acorn** is a nut.
It comes from an oak tree.
Squirrels like to eat acorns.

6 HOW PLANTS GROW

Plants grow in many different places.

Animals, wind, water, and people help to spread seeds.

This **maple** seed was carried by the wind.
It landed on the ground.
It began to grow when it had air, water,
and the right kind of soil.

People grow
many kinds
of plants.

People grow
plants
indoors.

People grow plants **outdoors.**

Plants are used to make our world more **beautiful.**

People grow plants to help keep the soil from washing away.

How do plants do this?

MAIN IDEAS

- Plants need air, soil, water, and light to live and grow.
- Most green plants have roots, stems, leaves, flowers, fruits, and seeds.
- Each part of the plant has a special job to help keep the plant alive and healthy.
- We eat many parts of a plant.
- Many plants grow from seeds.
- People use plants in many ways.

QUESTIONS

1. What does a plant need to stay alive and healthy?
2. What do the roots of a plant do?
3. What does the stem of a plant do?
4. What does the leaf of a plant do?
5. What does the flower make?
6. What is a nut?
7. How are seeds spread?
8. How do people use plants?

SOMETHING EXTRA

This is an aloe plant.

It can be used as medicine.

The liquid inside the leaf is used on cuts and burns.

The liquid helps the burns heal.

4 crayons = 38.6 grams

THINK METRIC

4 milliliters
of milk

5 centimeters

1 battery =
87.8 grams

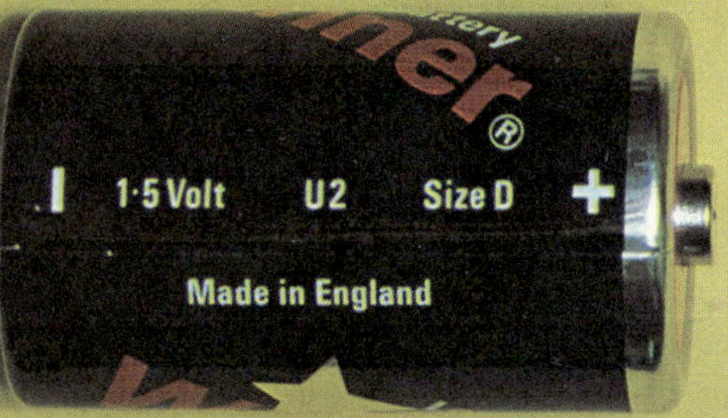

3 jacks = 5.5 grams

4 of these = about 1 liter

13.75 centimeters

PHOTO CREDITS

All HRW Photos by Ken Lax except the following:

Unit 1: p. viii–Girl Scouts of America; p. 1–Four by Five, Inc.; p. 2–*top* Rona Weissler-Tuccillo; p. 7–*bottom* Webb Photos; p. 8–*top* Four by Five, Inc.; *bottom* HRW Photo by Russell Dian; p. 12–*top* Michael and Barbara Reed/Animals, Animals; *bottom* Norman Myers/Bruce Coleman, Inc.; p. 16–*top left* Mira Atkison/DPI; p. 24–*bottom* Ken Lax; p. 25–*bottom left and right* HRW Photo by John Running.

Unit 2: p. 28–HRW Photo by John Running; p. 31–*top* © 1977 Imagery; *bottom* HRW Photo by John Running; p. 32–*top* Ed Nowak; p. 42–HRW Photo by John Running; p. 44–*right* Gerard Champlong/The Image Bank; p. 46–*top* HRW Photo by William Hubbell; p. 48–Grant Heilman.

Unit 3: p. 54–The Image Bank; p. 55–B. Golberg/The Image Bank; p. 56–*top* Four by Five, Inc.; *bottom* John Running; p. 57–*top* R. Gelberg/The Image Bank; *bottom* HRW Photo by John Running; p. 61–*top* HRW Photo by Russell Dian; p. 62–*top* Daniel Brody/Editorial Photocolor Archives; *bottom* Ken Lax; p. 63–*top* © Norman Tomalin/Bruce Coleman, Inc.; *bottom* Grant Heilman; p. 65–*top* Larry Dale Gordan/The Image Bank; p. 66–*top* HRW Photo by Russell Dian; p. 68–Ardea Photographics/London; p. 69–*top* Grant Heilman; *bottom* Mort Beebe/The Image Bank; p. 72–© 1977 Imagery; p. 73–Peter Arnold Photo Archives; p. 74–Ken Lax; p. 77–*top* Grant Heilman; *bottom* Grant Heilman; p. 78–*top* © Stouffer Productions/© Animals, Animals; *center* Grant Heilman; *bottom* W. Bayer/Bruce Coleman, Inc.

Unit 4: p. 82–Lee Foster/Bruce Coleman, Inc.; p. 83–© Zig Leszczynski/Animals, Animals; p. 86–*all* Grant Heilman; p. 88–*top* Grant Heilman; *bottom* © H. Wendler, 1976/The Image Bank; p. 89–*top* Peter Vadnai/Editorial Photocolor Archives; *bottom* © Dan McCoy/Rainbow; p. 92–*bottom* © Luis Villota/The Image Bank; p. 93–*top* © Luis Villota, 1973/The Image Bank; *center and bottom* Porter/Taurus Photos; p. 94–*bottom* Ward's Natural Science Establishment, Inc.; p. 97–Grant Heilman; p. 98–*bottom* Manfred Kage/Peter Arnold Photo Archives; p. 100–*top* William H. Amos/Bruce Coleman, Inc.; *center* © C. Newbert/Bruce Coleman, Inc.; *bottom* Grant Heilman; p. 101–*top* National Audubon Society/Photo Researchers, Inc.; *bottom* Pat Morris/Arden Photographics/London; p. 102–*top* Grant Heilman; *bottom* Ardea Photographics/London; p. 104–*top* Peter Arnold Photo Archives; *bottom* © Russ Kinne 1974/Photo Researchers, Inc.; p. 105–*top* Grant Heilman; p. 106–*bottom* P. H. Davis/Bruce Coleman, Inc.; p. 109–L. L. T. Rhodes/Animals, Animals.

Unit 5: p. 110–George K. Bryce/Animals, Animals; p. 111–Michael and Barbara Reed/Animals, Animals; p. 112–*top* © Carolyn Weatherly/The Image Bank; *bottom* © L. and D. Klein, 1978/Photo Researchers, Inc.; p. 113–*top* Hans Reinhard/Bruce Coleman, Inc.; *bottom left* Breck P. Kent/Imagery; *bottom right* Ardea Photographics/London; p. 114–*top* Ardea Photographics/London; *center* © Leonard Lee Rue III/Animals, Animals; *bottom* Four by Five, Inc.; p. 116–*top* Hans Reinhard/Bruce Coleman, Inc.; *bottom* Dr. Derringer/Imagery; p. 117–*top* Max Hunn, National Audubon Society/Photo Researchers, Inc.; *center* Alan Blank/Bruce Coleman, Inc.; *bottom* G. Ronald Austing/Bruce Coleman, Inc.; p. 118–*top* Und Berggen/Ardea Photographics/London; *bottom* Hans Reinhard/Bruce Coleman, Inc.; p. 120–*top* Jane Burton/Bruce Coleman, Inc.; *bottom* Four by Five, Inc.; p. 121–*top* Stan Keiser, The Sea Library/Photo Quest; *bottom* © Zig Leszczynski/Animals, Animals; p. 122–*top* A. Grotell; *bottom* © Dr. Wm. J. Jahoda, National Audubon Society/Photo Researchers, Inc.; p. 199–*top* Zig Leszczynski/Animals, Animals; p. 125–*bottom* F. E. Unverhau/Animals, Animals; p. 126–*top* Manuel Rodriquez; *bottom* S. C. Bisserot/Bruce Coleman, Inc.; p. 127–*top* Peter Arnold Photo Archives; p. 129–*top* © 1973 Jen and Des Bartlett/Bruce Coleman, Inc.; *bottom* Ardea Photographics/London; p. 130–*top* Imagery; *bottom left* Lynn M. Stone/Animals, Animals; *bottom right* Ardea Photographics/London; p. 131–*top and center* Imagery; *bottom* D. Hughes/Bruce Coleman, Inc.; p. 132–HRW Photo by Russell Dian; p. 137–Imagery.

Unit 6: p. 138–C. E. Schmidt/Taurus Photos; p. 139–Elyse Lewin/The Image Bank; p. 140–Webb Photos; p. 142–*top* Jay Lurie; *bottom* Rona Weissler-Tuccillo; p. 145–*top and center* Webb Photos; p. 148–*top* Richard G. Langer/Taurus Photos; p. 149–Webb Photos; p. 152–Webb Photos; p. 153–*top* Christina Wohler; p. 156–*top* Four by Five, Inc.; p. 159–*top* Webb Photos; *bottom* W. H. Hodge/Peter Arnold Photo Archives; p. 161–Bill Swan; p. 165–Webb Photos.